An introduction to GEOLOGY in the PEAK NATIONAL PARK

GEOLOGY affects us all.

The raw materials we need,
the scenery we enjoy,
the places where we can build our homes,
our roads and railways,
our water and food supplies—
all these depend on geology.

Cover: Alport Castles—a classic landslip in the north of the National Park.

THE PEAK NATIONAL PARK

In the Peak National Park it is the variety of natural beauty that strikes you first. The high, heather and peat-covered moorlands of Kinder Scout and Bleaklow, with dark peat haggs and gaunt, bare, rocky crags, contrast strongly with the White Peak landscape of drystone walls, green patchworks of fields and woods, and deep precipitous valleys. How did this come to be? What is there to look for? This booklet aims to help you find out more about the ways geology has shaped the Peak, so you can understand more about what you see on your visit. Geology is a wide subject, and this introduction cannot be a textbook nor explore all aspects of geology, so we have listed other books inside the back cover if you would like to read more.

Perhaps you know about Blue John, the purple fluorspar found near Castleton? But did you know that millions of years ago some of the present-day Peak District was coral reef of the kind that surrounds tropical islands? To explain these things we need to take a look at the story of the Peak's rocks.

The White Peak—a landscape of limestone walls and green fields, with scattered farms, villages, and small woodlands. The "bare bones" of the whitish limestone rock stick out through the soil, and form outcrops on the valley sides.

The sombre moors and gritstone outcrops of the Dark Peak. The hard gritstone forms a horseshoe across the north, and down the two flanks of the Peak National Park. Here the more gentle peat and heather covered slopes give way abruptly to the precipitous edges.

The softer shale is often eroded to form river valleys. Where these are sheltered, hedges sometimes provide a contrast to the more usual drystone walls.

THE STORY OF THE ROCKS

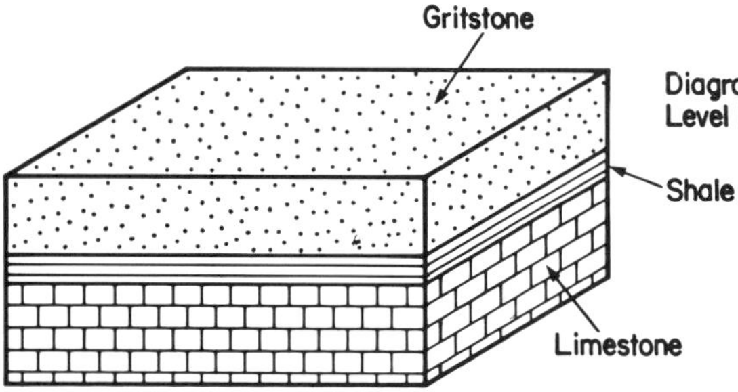

Diagram 1
Level layers

Imagine a sandwich cake made of rock. But the bottom layer of cake is **limestone** (its full name is Carboniferous Limestone, meaning limestone made in the same period as coal, about 300 million years ago). The jam is the middle is **shale,** a flaky hardened mud, and the top layer of cake is **gritstone** (Millstone Grit) a coarse, hard, sandstone.

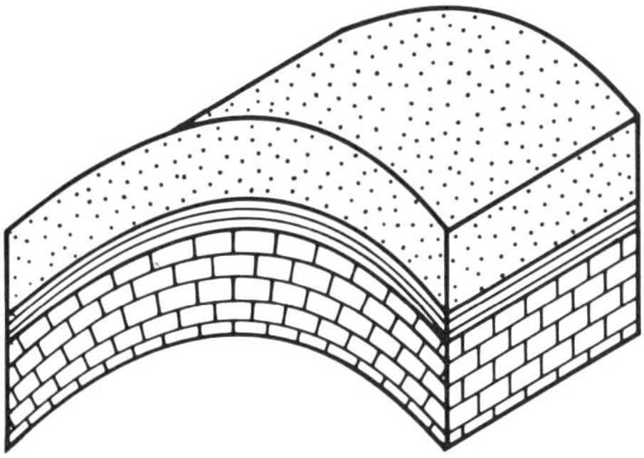

Diagram 2
The layers
become folded

Then imagine bending the cake, and slicing the top straight off. Look at the diagram and you can see how the layers of rocks, like the cake, became bent by millions of years' pressures and forces acting upon them.

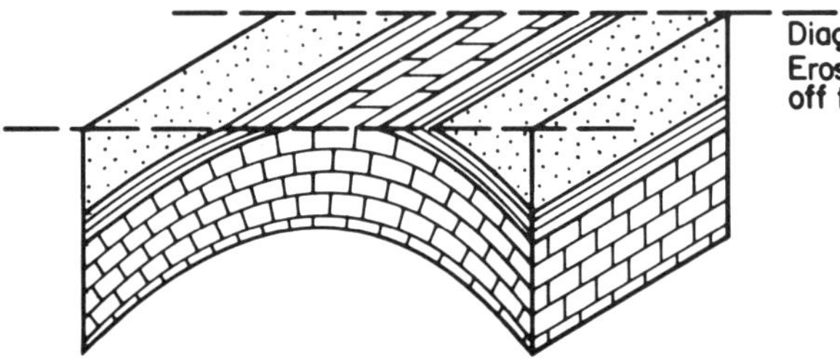

Diagram 3
Erosion levels
off the top

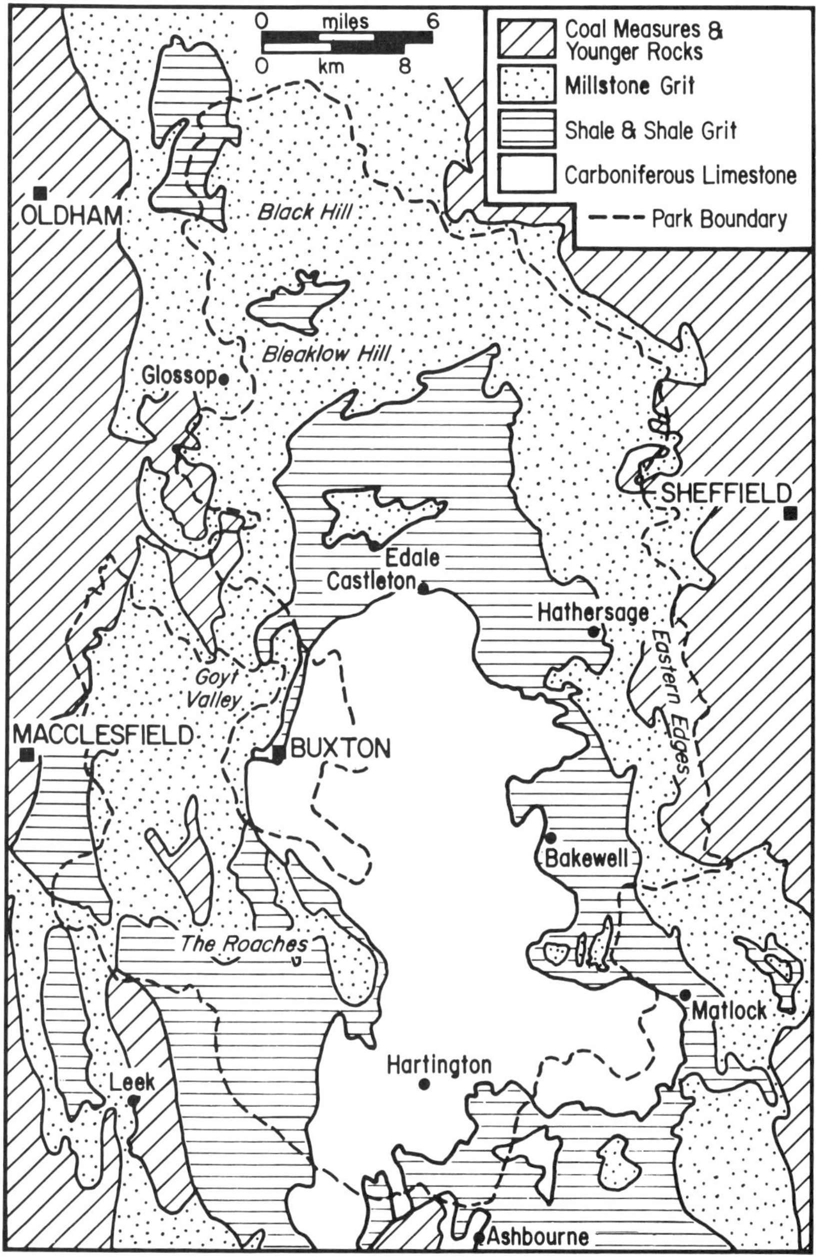
0 miles 6
0 km 8
Coal Measures & Younger Rocks
Millstone Grit
Shale & Shale Grit
Carboniferous Limestone
Park Boundary
OLDHAM
Black Hill
Bleaklow Hill
Glossop
SHEFFIELD
Edale
Castleton
Hathersage
Eastern Edges
Goyt Valley
MACCLESFIELD
BUXTON
Bakewell
The Roaches
Matlock
Hartington
Leek
Ashbourne

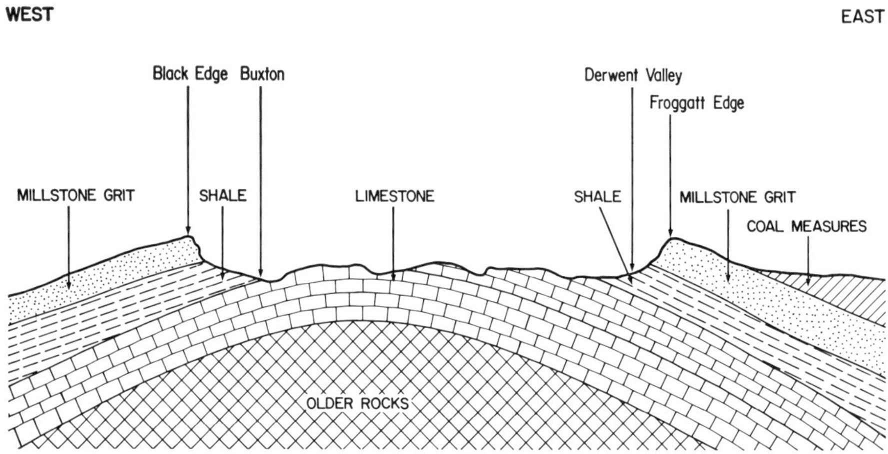

The bending was done by forces in the earth pushing the rocks into gigantic ripples. But some rocks are harder than others. Our limestone is fairly hard, the shale is softer, while the gritstone is the hardest of the three, so the surface exposed by our "slice across the cake" is worn away at different speeds by the action of rain, rivers, wind and ice—the forces of "erosion". The shale is softest and wears away quicker than limestone, often making valleys. Gritstone being the hardest, it is this rock which makes the high moors and cliff-like "edges" as they are called in the Peak District. The limestone has eroded into a flattish plateau. Deep gorges in it were cut by rivers as the land slowly rose during millions of years time.

It isn't quite as simple as that—the "folds" in the rocks are more complex and where the layers of rock have fractured, there are "faults".

300 million years ago the climate was different. Corals grew under a warm tropical sea, and for millions of years the shells and remains of myriads of sea creatures built up on the sea bed. This hardened into limestone rock, and you can see the layers of fossil shells to this day. The grey or white limestone (that's how we get the name White Peak), may be in thick massive layers (called strata). Sometimes it is in thinner layers, according to the conditions in the sea when it was formed. At the edges of the sea coral reefs grew up in places, with shells in amongst them. When turned into rock, these stood up better to the forces of erosion and can today be seen forming prominent hills in the Dove Valley—Thorpe Cloud, Bunster, Chrome, and Parkhouse Hills, and around the Winnats.

Typical "edges"—gritstone escarpments near Curbar.

Limestone reef knolls stand up to erosion.

As the limestone was eroded, spectacular features developed—the spires and pinnacles of the dales, and the caves that are found in much of the area. Limestone is dissolved by slightly acidic rainwater, so cracks and joints were gradually widened into caves. Castleton is well known for its caves, and there are several you can visit. Another feature of limestone country is the "dry" valleys (e.g. the Winnats Pass), where the water runs through cavities below the surface of the land. In other cases streams and rivers "disappear" in dry weather as the reduced flow is beneath the normal bed, for example the River Manifold which re-emerges in the grounds of Ilam Hall, where it leaves the limestone.

Although the soil is fairly good on the limestone, there is a high rainfall, which washes out much of the lime, so you may see the farmers actually putting lime on the land in these areas. When there were no piped water supplies, the siting of homes and entire communities depended on a good and reliable supply of water from a spring, or well. In many villages on the limestone, where this applied, a thanksgiving in the form of the traditional well-dressing is still carried on.

Caves are a typical feature of limestone areas.

A limestone dale—the dry valley cut deep into the plateau.

Geology determines water supplies—a dewpond for cattle in the White Peak limestone country.

The "jam" in our sandwich cake is shale. This was made by fine silt from a great river to the north-east settling on top of the limestone. Nearer the mouth of the river the heavier pebbles and grains of sand sank first, forming gritstone. The position of the coastline and also the level of the land changed continuously, so did the areas where these gritty sandstones (gritstones) were dropped. So we get layers of shale and grit building up. Where it was mainly shale we can see it was usually softer—often eroding into river valleys. Landslips are often found, too, the sandstone slipping on the shale and spilling down the hill in hummocky masses—for example at Mam Tor, and in part of the Edale Valley.

The solid gritsone is much harder and gives us the "edges"—characteristic of the eastern part of the Peak in particular, for instance Stanage, Froggatt and Curbar Edges, making good crags for rock climbers. In the west we have Axe Edge, and the Roaches. Gritstone gives us poorer agriculture, with acid soils. It does not let water pass through it as the limestone does, so boggy areas tend to build up, often with peat and heather on top.

The "face" of Mam Tor near Castleton—strata exposed by landslip. Note the hummocky slipped material below and the Losehill ridge in the background.

Landslips are a force to be reckoned with—the damaged road by Mam Tor.

There are very few fossils in these rocks—just a few poorly preserved shell-fish remains in the shales. The swampy delta conditions when the next layers of rock were being formed, helped them to support some vegetation, and so the coal measures eventually came into being. There is an area of coal measure rock in the Goyt Valley, with traces of the coal which has been mined there.

Limestone, shale, and gritstone are all **sedimentary** rocks. These are derived from other rocks which over a long span of time have been gradually broken down into small particles, and re-deposited in layers. **Igneous** rocks have cooled from a molten state. There are comparatively few visible in the Peak area, but two which do exist are dolerite and basalt, formed by small volcanoes which erupted through the carboniferous sea. The third kind of rock is called **metamorphic,** which means it has had its shape and structure changed by intense heat. There are one or two areas of metamorphism in the Peak, where molten igneous rocks have changed the structure of the rocks with which they came into contact, for example on the approach to the old quarry in Tideswell Dale, and a few hundred yards south of Peak Forest Church.

FOSSILS

Britain's climate is a favourite conversation piece. We usually grumble because it's too cold or wet. But if you could cast your mind back in time and see the area as it was millions of years ago, it would be a very different story. For in Carboniferous times—about 300 million years ago—much of what is now the Peak was covered with a shallow, warm, tropical sea. Measurements of ancient magnetic minerals indicate that Britain then lay close to the equator, and has been drifting northwards ever since. Here and there small coral atolls were growing up, with reefs around them teeming with coral and other marine life. The reefs eventually were buried and fossilised, only to be exhumed in the comparatively recent past. Some can be seen today as the reef limestones, notably near Castleton and in parts of the Dove Valley. In the tropical seas, as small creatures died, their shells fell to the sea floor to be broken down by the action of tides to form eventually the thick-bedded limestone of the White Peak. From time to time you can find layers of fossils in the limestone which show where the current brought together a great mass of shells.

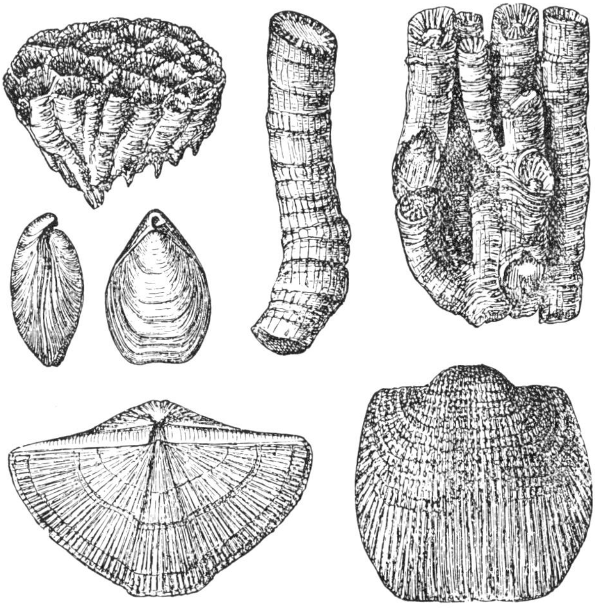

Typical Carboniferous fossil corals and shells—from an old engraving.

A sea lily, or crinoid

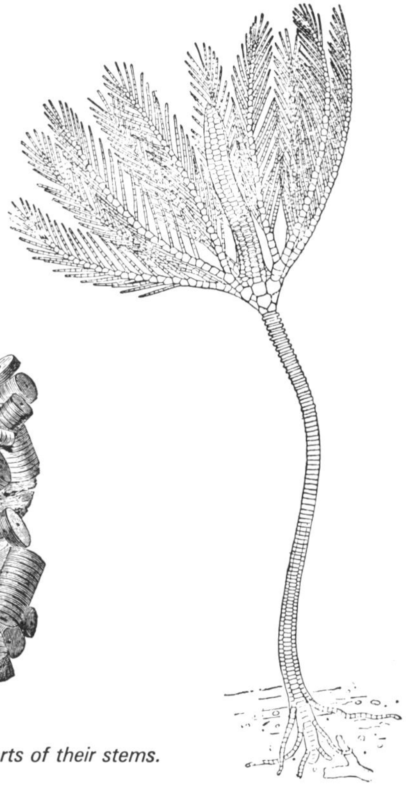

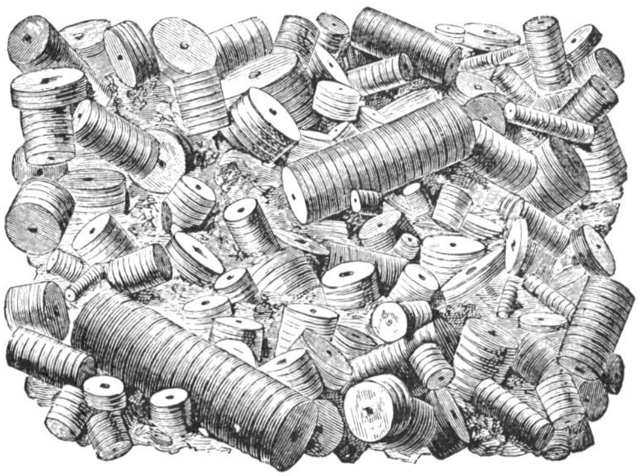

—you can often find the fossilised screw-like parts of their stems.

The study of fossils (known as palaeontology) would make a book in itself. Although the millstone grit and shale in the Peak will yield some fossils (mainly shells similar to present-day mussels, and some fragments of trees), it is the limestone which provides an exciting variety of fossils. From place to place on the floor of the shallow, warm sea of the time were forests of "sea lilies", not plants but in fact animals related to the star fishes. These sea lilies could grow up to a height of 8 or 10 feet, with a stalk made up of segments and the main body at the top with five branching arms. When they died, the segments fell to the sea bed, and millions of these can be seen to this day in fossil form—individual segments looking like beads, while some have a screw thread appearance and are known locally as "Derbyshire screws". Around these fixed creatures were millions of small animals swimming in the sea, or crawling on the mud, in particular the lamp shells, and interesting molluscs, among them the goniatites, with beautifully coiled shells.

ECONOMIC GEOLOGY

The millstone boundary markers which you pass on the main roads coming into the Peak National Park are reminders of the economic significance of the rocks and minerals found here. At one time used in local industry, and even exported, partly-worked millstones can still today be found on some of the gritstone edges. Traditionally, Peak District buildings are of local stone—limestone, with the harder gritstone

Peak District buildings traditionally used local stone.

Quarries can make a significant impact on the landscape.

used for lintels, window surrounds, and so on. Some gritstone continues to be worked for building stone.

Limestone forms the raw material for a wide variety of products, and in parts of the Peak is up to 98% pure calcium carbonate, making it one of the major sources of this basic industrial chemical. So it can be used in the making of such diverse products as glass, paper, petrol, sugar, paint and toothpaste. Its chemical properties are also used to remove impurities during the manufacture of iron and steel. Two other uses of the chemicals in limestone are for agriculture, to reduce soil acidity and in the manufacture of cement, which is carried out in the Hope Valley—this industry using the chemical properties of the shale deposits as well. The deep, even layers of limestone in the area make it comparatively easy to work, and the closeness of the National Park to major centres of population means that there is a great demand for limestone as an aggregate in road and building construction. In fact, the rate at which it is being worked for this purpose is causing concern to the Peak National Park Authority, particularly in view of the scale of the workings, and the need to conserve an important raw material for the chemical industry.

When the beds of limestone were folded after the Carboniferous period (remember our diagram on Page 4), they formed a dome, rather than a simple fold. As this happened, cracks appeared in the limestone running roughly east to west. What then appears to have happened is that minerals crystallised out from hot solutions originating deep below the earth's surface, flowing through the cracks. The shale above seems to have acted as a cap rock to prevent them escaping further, so they settled in different bands (veins) according to the different temperatures at which they crystallised. This means that different minerals are found at various points along a vein.

Lead ore (galena—or lead sulphide) occurs widely throughout the veins, and has been worked since Roman times—an industry which reached its peak in the 18th and 19th centuries. But barytes (barium sulphate—sometimes called heavy spar) is found mostly in the central section, and fluorspar (calcium fluoride) mainly occurs in the east. Barytes is used in making some chemicals, as a filler in rubber and paint, and as a weighting agent in the drilling muds used in oil and gas fields. There are two main uses of fluorspar—as a flux in the steel industry (this is called metallurgical grade fluorspar), and as a source of hydrofluoric acid which is widely used (this is known as acid grade fluorspar), including the production of aerosols and refrigerants, and in the manufacture of aluminium. There is a third grade of fluorspar (ceramic grade), which is

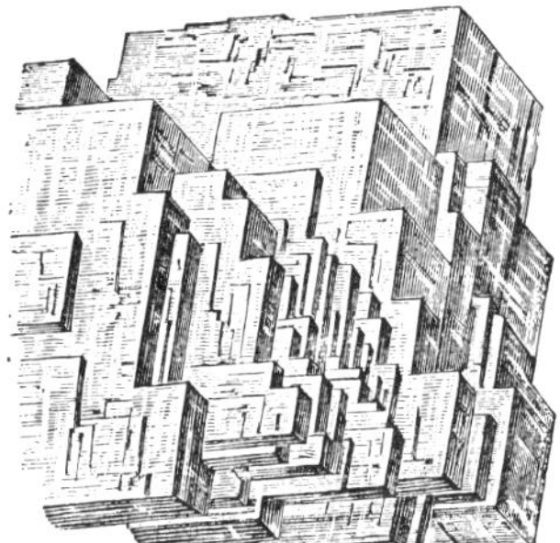

Calcite.

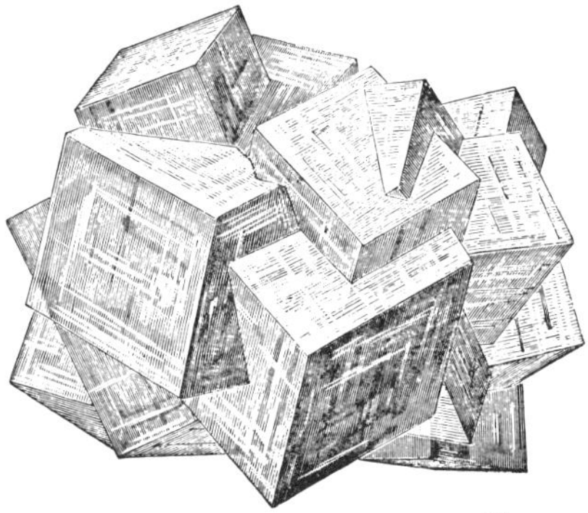

Fluorspar.

From an old engraving.

Deserted mine buildings, and disturbed ground tell the story of our search for lead. Magpie Mine.

used in making glazes and enamels. The great majority of Britain's fluorspar comes from the Peak, and although much can be worked underground, and surface workings restored, there can be unsightly problems caused after processing by disposing of much waste slurry in large artificial lagoons.

After the Carboniferous period when the Millstone Grit was worn away and the limestone again became exposed, deposits of sands and clays were formed in depressions in the limestone. They are called "pocket silicas" and are worked in various areas, for example near Friden, for making into refractory materials—firebricks, and furnace linings for industry. Some calcite is also worked in the Peak, mainly for use as a decorative finish for buildings. It is also used in making white paint for road markings. In addition to these minerals, a little zinc has been found, and copper was formerly worked, notably at Ecton in the Manifold Valley.

During the 18th and 19th centuries, black Ashford marble, which is really a dark, hard limestone, was inlaid with patterns and worked into ornaments and jewellery, but probably the most well known ornamental mineral in the Peak is the famous Blue John. This mineral, found near Castleton, is a purple-blue variety of fluorspar, often found in attractively coloured bands. There are several caves in which you can see the mineral and it is made into jewellery and ornaments for local sale.

PLEASE HELP!

Geology is now a popular subject and everyone who studies it seems to want to wield a hammer and bash away to get samples! But in some parts of the National Park (and elsewhere), important geological sites have been destroyed by over-enthusiastic collectors, and are now valueless for scientific study. Farmers and landowners have complained of trespassing, and also that walls have been damaged, in the search for specimens. So **please** make sure you have permission to be on the land in the first place, and don't hammer **anywhere** indiscriminately. Collecting specimens is often unnecessary, but if you **must** find something to bring back with you, try looking over waste tips from former mining activities, or take photographs instead. Your co-operation will make life easier for others who follow, and for the local people as well. Please follow the Country Code at all times. The Geologists Association produces a "Code for Geological Fieldwork" in the form of a leaflet containing useful advice. The Association's address is c/o The Geology Department, University College London, Gower Street, London, WC1E 6BT.

SOME WORDS OF WARNING

Geology takes us into many fascinating kinds of landscape—to rock faces, landslips, mines, and quarries, and perhaps areas with abandoned and overgrown mine shafts. Please take care—unnecessary risks may lead to accidents and put the lives of others at risk too.

FURTHER ADVICE

If you want to find out more about geology, visit your library and museum. Many evening classes on the subject are also run, and your Adult Education Centre may be able to help. Local natural history societies often have members who are interested in geology, and by joining them you could enjoy well organised field visits, and learn about rocks, minerals and fossils at first hand.

BOOK LIST

Title	Author	Publisher
The Peak District	Edwards, K. C.	Collins
The Peak District	Millward & Robinson	Eyre Methuen
Geology Explained in the Peak District	Wolverson Cope, F.	David and Charles
Geological Excursions in the Sheffield Region and the Peak National Park	Neves, and Downie (Editors)	University of Sheffield
Nottingham and its Region	Edwards, K. C. (Editor)	British Association for the Advancement of Science
Sheffield and its Region	Linton, D. (Editor)	British Association for the Advancement of Science
The Caves of Derbyshire	Ford, T. D.	Dalesman
Limestone and Caves of the Peak District	Ford, T. D. (Editor)	Geo. Abstracts Ltd., Norwich
Geology of the Country around Chapel-en-le-Frith	Stevenson and Gaunt (Institute of Geological Sciences)	H.M.S.O.
Map of Castleton—Geological Sheet SK18	—	Ordnance Survey

There are many other excellent books written as an introduction to the subject of geology, some specifically for young people. Your library or bookshop should be able to advise you.